Re-Engineered Economy

Low Power Banks

High Performance for Consumers

By Edward Seymour (PoemsOnTheSpot)

Re-Engineered Economy

Low Power Banks

High Performance for Consumers

ISBN: 978-1-304-61275-5

Preface

This text is dedicated to defining a new economic model that can help change the course of our future. It is written in the form of a series of mathematical proofs showing how current methods of measuring the health of our economy are fundamentally flawed and need to change.

The changes proposed within this text are based on principles used to recover from the "Great Depression" and apply to our contemporary scenario equally. The crash of 2008 in many ways served us a blow as devastating as 1929 but media spin doctors are unwilling to agree.

The perspective used to devise these changes is the study of Engineering and Business Finance and each of these proposed changed delineated within this text can stand the scrutiny of business case analysis. My intention in creating this volume is to offer insight into fairly simple changes that fit within current banking laws and would be relatively simple to implement. I openly welcome a national debate thereof.

Along the way I also suggest some changes to banking policy and practices that would help bolster our economy in ways that tilt the balance in the direction of those who have less to start with.

Table of Contents

About The Author/Poet

I have studied Engineering and Business Finance and have had the pleasure of a long standing career within a computing giant known for its history of supplying reliable, predictable computing platforms to Banking and other large industries. It is within this context that I learned to implement business cases to acquire funds for projects and deliver on aggressive schedules within severe constraints.

In the course of my work, I have often challenged the norm and questioned prevailing methods. Some would call this thinking outside the box. As it applies to our current economic model, there are some fundamental issues with core principles that need to be challenged to effect this plan. My work within the computer industry has afforded me insight that I share within this text.

Mythical Measures Of Financial Concern

To effect fundamental change in our economy is to directly contend with some of the contemporary ways in which banking and finance are conducted. I will in this chapter dispel the myths upon which these current measures are built and propose concrete alternatives.

These concepts are a result of studying banking as a consumer and interactions with countless professionals in the field of finance. They have been validated as well by the publication and limited distribution of a survey document in 2010.

Credit Score – The Myth

Background -

This is a term is not one that has an extended history and was in fact invented during the 1980s as a measure of who deserved to get a credit card or a loan. The measure is currently viewed as a golden measure of "credit worthiness" and is illustrated in the figure below from http://www.philly.com/philly/classifieds/cars/research/general_cars/General_History_of_Credit_Scores.html, http://www.ehow.com/about_4613002_how-did-credit-scoring-come.html

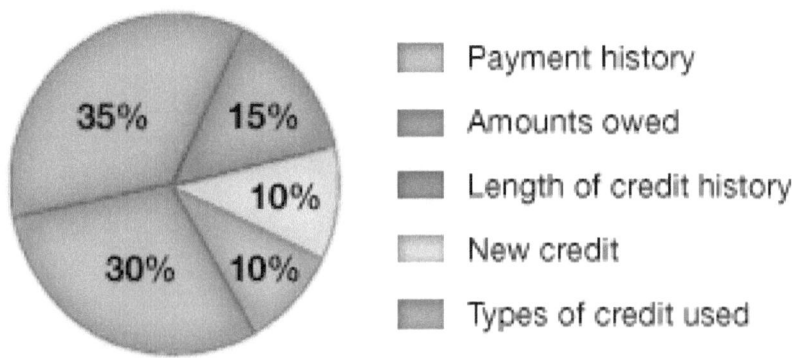

This credit score concept was the intellectual property of a corporation known as "Fair Isaac Corporation in 1958" but came into wide spread use in the early 1980s and even more in the 1990s for mortgages. It is commonly referred to as FICO and deemed to be reliable in that it includes numbers from three different corporations in the calculations.

If you dig into the calculations you will find a disturbing trend. If you were a bank trying to determine "credit worthiness" of a client, details on the items included in the credit score should be taken into consideration but as an equal counterbalance to a savings score.

Truth

As it stands, one can attain a perfect credit score for acquiring the maximum potential indebtedness as long as you make minimum payments on time and use all lines of credit to the tune of 20%. In effect this implies that you have hyper extended your income to the point where you will find yourself bankrupt with the slightest change.

The scoring system also severely penalizes anyone who pays predominantly in cash or debit cards, noting them as unworthy of credit. In extreme examples this also adds a flag to indicate that you are likely money laundering.

In Chaper 2, I will propose a Savings Score to be used as a counter balance to Credit Score which will result in accurate measures of financial risk. One measure in the absence of the other assures failure.

In Terms Of Gambling

To have a perfect credit score, you are a "perfect mark" You provide a constant income for credit card companies as follows:

To maintain the perfect score you keep 20% of your credit always used. If your credit limit is $5000, with perfect credit score you get an annual percentage rate of prime + 9 % ~ 9.25%. This translates to (.0925/12*.2*5000) per month as a base fixed income to them ($7.70) plus a tax of 3% on everything you buy (assume $5000-1000 balance you always carry)*3% = 4000*.03 which yields $120+7.70 for 4000 lended for a max of 28 days.

Net profit to credit card company $127.70 on 4000 of float

9

Real Interest Earned By Them

Amount Received / Capitol Extended Continuously (X12)

$(127.70/4000) \times 12 = 38.31\%$

Sustainable Alternative – Savings Score

Consumer Based Economy - Myth

Background

This is another commonly used term that reflects fundamentally unsound measures. To judge the strength of our US Economy solely on how much we collectively spend is flawed. There used to be GNP (Gross National Product) and that was a measure of how much we produced.

Other economic measures were then compared to the total production from our country. Then there were other measures of Trade that showed how much we exported versus how much we imported

In a country that still has as its most valuable possession almost unbounded resources and raw materials, we should once again return to measuring how well we use our resources, not how much we buy from others while giving our raw materials away for a song.

In Terms Of Gambling

We as a country buy into the idea that as long as we buy things, we make the country healthy. Implied in this statement is the idea that to buy is being a loyal American where what we buy is largely products produced off shore.

Tie to gambling is the idea that a small wager you place on a horse will yield many times over your "investment" IF you pick the correct horse.

Net Effect

As we buy things that are produced elsewhere at a cheap price we ensure that jobs associated with these goods and services will continue to be supported elsewhere at a much higher net cost.

When we overextend ourselves to keep the economy afloat, we, in effect perpetuate the cycle.

Alternative Answer – Reward Savings

This is the sustainable answer to the problem that was in play following the depression of 1929. A measure of financial health was to demonstrate the confidence We The People had regained in banking as defined by Savings And Loan.

Cash Is A Thing Of The Past – The Myth

If your perspective is to increase your credit score, stay far from using cash in any transaction and avoid using a checkbook or debit card as each use dings your credit. Banks would like to see this trend continue to the point that you would be unable to identify currency as we know it.

One of the three principals of credit score calculations is the propensity you have for paying for things in cash. Too much use of cash means you are

1. Not able to secure credit

2. A criminal who does not want funds traced

3. Too stupid to leverage your income

News flash, just in:

1. Use of credit or debit cards guarantees banks take a cut of each transaction before merchants or governments

2. Spending trends are tracked and sold

Chapter 4 will be dedicated to how Cash is the four letter word we need to pronounce clearly and use often. This has direct ties to the Savings Score aforementioned and covered in Chapter 2.

In Terms Of Gambling

If consumers use Cash to pay for things, the cut available to credit shark card companies is zitch, nada, zero. This, my friends, is a problem of national importance, that ensures credit scores of any such ill-informed consumer should suffer badly

Net

If you are so stupid as to use Cash for transactions, you will become cursed as you try to buy a car, buy a house or apply for a credit card. How could you be so inane to deny yourself access to platinum cards, chock full of rewards.

Real Alternative

Pay in Cash. Even have the gall to carry bills that are not typically dispensed by ATMS and that seldom used form of payment called COIN.

No Loan Is Ever To Be Paid Off, Just Refinanced Or Written Off – The Myth

Background

This mantra is so prevalent in our economy it is nearly accepted as an axiom. It should be considered a myth for reasons of sustained economic growth. This is akin to the principle of placing saving rate above all.

Based on sound financial principles, when you acquire a loan you agree to repayment terms which result in a mutual understanding of obligations but typically come with the statement that there is no prepayment penalty which implies that early repayment results in much less interest.

From a pure financial standpoint early repayment results in a premature cash flow to the lending institution which is in effect converting a standard loan into one including a balloon payment where they receive a large influx on cash when you pay off.

A simple comparison of holding a loan for 30 years to that of paying it off in 5 years will show there is an enormous difference between the two methods that is not currently taken into account for mortgages, especially those that are refinanced.

In Terms Of Gambling

For this point, I will focus purely on home mortgage operations. You "buy" a house by entering into a agreement to make payments. These payments are in the form of Principle, Interest, Taxes and Insurance along with Interest Payments On Points Of Origination and various other fees. The agreement is set for N years of repayment on a schedule like a loan.

The gamble is as follows. You buy the house for price X thinking that the price you paid was a good investment. Furthermore, you wager that your ability to repay this loan will not change meaning your job will continue as it has.

Traditionally the purchase of a home would also imply that you would remain in the one residence for the duration of the loan and reap the benefit of no house payment after that period. Enter into the realm of gambling brokers in the field of "real estate" and the picture changes.

You now decide which how to buy based on profit you make by selling this house as its value increases. To avoid taxes you always buy something more expensive that the one you sold. This often yields the behavior of sign up for the most expensive house you can as you will simply sell and move up. This ensures that the payments you make toward the home that you "buy" are allocated to interest on points, taxes and insurance that are paid from escrow and little is ever applied to pripical.

The yield of this gamble pays handsomely to the brokers. The risk is yours to bear.

Net Effect

As we gain ownership in a home, the impetus is there to "use the equity" to pay for unexpected expenses which means you refinance and roll back in the costs of doing so. This ensures you continue to pay interest on interest.

Real Alternative

When you buy a home, examine the amortization table and reduce the term of the loan by doubling principal payments. Be sure to pay with checks so you have a defined paper trail. When you reach a 20% equity position, demand that you pay taxes and insurance yourself (no more escrow) which will have an impact on how fast principal can be paid. Any excess held in escrow is money that should be applied to principal but the brokers now have decided that they need to lower their risk by padding required funds and keeping the float on your money.

Rewards Cards And Programs Are Good For You

Background

Conceptually when you hear the word Reward, the idea that comes to mind is a well deserved gift or prize. They lure people into such programs by stating they will offer you 1% or even 5% back on certain purchases. These offers are another form on myth in that they are never free. Even the airline cards which offer "free" tickets charge a fee to redeem them and often charge an annual fee for enrollment.

These gimmicks are used to entice many people into credit card programs with very high interest rates in exchange for "free" things. The mere use of credit cards and debit cards for all purchases ensures the provider of the associated credit gets to charge their fee for each and everything you purchase. This results in a cost of doing business for each and every merchant that is considerably higher than cash transactions.

The result thereof is the merchant needs to charge you a premium price for goods or services that could be much lower with cash transactions. Talk to any merchant, the cost of using any form of plastic is multifaceded.

- Fee paid to the card provider
- Fee paid to "Square" or other service
- Fee paid to simply connect to the Internet
- Fee paid to Pay Pal

In Terms Of Gambling

Rewards programs "pay" in restricted funds. Not really different than trading your real cash for poker chips or coins for tokens at a video game place. Even the 1% cash back has a required hold on payout. You must use the card for 6 months to a year before any rewards can be redeemed. At that point, you are offered higher value in redemption for using your prize to purchase products from a company that is affiliated with the card company. In essence you are a captive audience receiving a hard sell on these things.

If during your time of earning rewards, anything happens and a payment is late or you close the card and move elsewhere, rewards vanish. The gamble is yours, not theirs.

Real Alternative

Consider each card you have a balance on as a saving account that earns interest. By paying it off, you essentially pay yourself interest that is a real reward.

Save For The Future Is So Passé

Background

Banks will tell you that the era of passbook savings is forever gone, a thing of the past. They will note the current prime rate as evidence that for them to offer 5% interest on a saving account or even more on a certificate of deposit would evoke upon them financial ruin. This is one of the most blatant Myths (outright lies) that has ever been cited.

The core reason for not offering passbook saving is the impression held by many bank executives that savings based banking is not where the BIG money sits. Truth to be known it is far more profitable for them to wait for someone to overextend their credit and pounce all over them with fees.

A consumer who uses plastic for all transactions also tends to not see the same effect as opening a wallet to see if funds exist for a purchase or reconciling it against a budget. Au Contraire, transactions take place, funds reconciled where the probability of over consumption is vastly increased, "stimulating the economy".

In Terms Of Gambling

Each time you pay with credit, especially the resolving type, you offer the Casinos (banks) their take. The flat tax of instant 3% equates to 3% each month or 36% for the year. You also wager that your will have the money to repay the bet you made by the due date. Remember, if you are a compliant gambler, you keep 20% on the tab always which ensures that the Casino, keeps the lights on. You agree to a fixed cost of 20% multiplied by the "great interest rate of 10 percent times your credit limit.

Sustainable Alternative

Look at each credit card or line of credit as a savings account that pays you the interest as you eliminate a balance on the card. This is exactly what the Casino would like you never to do as it "costs them, not you"

ATM Fees Are A Normal Part Of Doing Business - The Myth

Banks will extol the virtues 24/7 banking as the benefit you glean from using an ATM. Chances are most people use them routinely for getting cash and checking balances. Truth to be known, they are vastly less expensive than the same bank having actual people available for such transactions.

In reality, each and every ATM becomes a CASH COW for the institution very quickly after installation. Cost of procurement is low in the range of under $1500 per machine and once installed, people pay up to $3-$5 per transaction to use them.

In Terms Of Gambling

Look at an ATM as a simple one armed bandit. You pull the lever as you insert your card and enter your pin while you are under surveillance. ATM Charges are the fee you pay to play with your money. Occasionally you win the $3 back by use of "in network" machine just like once in a few hours the one armed bandit "pays out"

Sustainable Alternative

Demand that all ATM transactions be free to use, anywhere. Demand that all machines be reconciled as frequently as teller shifts. This reduces risk to you and the bank of transactions gone awry,

Electronic Transfers Of Money Take Forever – The Myth

To pay a bill electronically these days involves a transaction that takes milliseconds to seconds to complete even from one bank to another or a bank to any other business. The advent of 24/7 banking has assured this to be true, yet banks and banking institutions insist on having your funds disappear for a matter of days in the process while control is passed from one "handler" of your money to another.

The simplest examples of this are Bill Payer systems and ATMs. Both have immediate liquidity of your funds. Both have immediate accounting for consumption of your money and the potential for immediate deposits but intentionally hold your funds for a while determined by some arbitrary means.

Some will claim these are regulatory rules by which they comply. If that is really true, the rules were set during the pony express times and need to be changed to reflect contemporary delays that are less than a second in most cases in any transfer.

In Terms Of Gambling

Electronic transfers of money with arbitrary and deceptive delays is the equivalent of trading your money for poker chips where you bet on the idea that the money you used to fund the transaction will arrive on time so you are not called late.

Sustainable Alternative

Represent that actual delays in electronic transfer of money from one computer to another in milliseconds it actually takes, not days or weeks currently recorded. Take the float out of cash based transactions.

Savings Basis Of Financial Worthiness – A Truth

This is not a new idea in concept as it was at the core of our recovery from the crash of 1929. Due to the run on banks people were distrustful of banks being trusted with funds and thus Savings and Loans were established with a federal guarantee of up to $100,000 under FDIC.

Conceptually cash deposited was viewed as working capitol which then was used to fund loans and pay passbook savings interest. The basic idea was banks would lend money at a rate a point or two above the rate they paid for savings. People were also rewarded for keeping savings in the bank for a prescribed period under the auspices of Certificate of Deposit.

Basic Debt Ratio Math For Lending – A Time Honored Truth

This method of evaluating a prospective customer for lending was essentially abandoned in the 1980's when the numbers would not support lending to enough people. Core principle was and remains sound. Essentially you look critically at your income sources and obligations. In essence, no more than 25-28% of your income should ever be allocated for housing of any kind and you should never carry a debt burden in excess of 36% of your income to insure you can repay the obligations.

The calculations are dependent on the interest rates offered but under these guidelines, no one would ever get into an obligation that would result in the level of foreclosure we are commonly seeing today. These ratios still make an enormous amount of sense and should be reinstated as a means for determining lending. Only the attribute of credit score that shows payment history should be ever used as the other elements that lower a credit score based on extensive use of cash, should be viewed as more positive, the lower the score.

Saving Score – A Way Toward Sustainable Growth

The best measure of one's financial sense is the percentage of income saved. Here again, a return to former measures of sanity should prevail. We as a country should be saving in a tiered fashion as follows:

1. 10 percent of your income should be saved in long term savings in the form of Certificates of Deposit paying 7-10% interest

2. Each person should have one year salary saved to account for unexpected expenses or economic changes and that should be paid 5% interest annually

3. All children up to age 18 should be paid 5-7% based on durations of savings for keeping money with the bank long term so that they have funds to help pay for college or trade school.

Banks will argue that the rates mentioned above are not viable with "prime rate" at an unprecedented low value of ~0%. Truth to be known almost no lending ever has much to do with that rate.

Banks pass money between them overnight for free. We sign up for "lotto based credit cards" with a lure of 0% interest only to see the real cost to us is 19% interest the moment the post office sneezes and your payment is late by one hour.

Even when all payments are made on time, each and every transaction using credit card, debit card results in a fee charged to the merchant which results in higher costs for them. Higher costs for them is reflected in higher prices for you.

Another, largely illegal practice, is the one that evokes a credit check whenever you pay for items using a debit card. The fact of the matter is that the business only needs to see that you can afford to overrun your expense by some margin. This results in a temporary hold on the funds in a credit card.

While the same exact procedure should apply for debit cards, the system instead runs a credit check with each use of a debit card for things like car rental. The effect is cumulative so that it ensures you degrade your credit by paying up front with cash as opposed to the higher margin form of payment (credit). This also gives reasons for the credit you receive to be at a higher interest rate due to your unwillingness to use credit.

In a savings score, you would be rewarded for paying bills with cash as it shows you are much less likely to get behind on a loan payment. In the same sense that you are dinged for any late payment in a period of 7 years, you should be rewarded in points for each day you pay a bill early before the due date. In the current credit score system a single late payment is the equivalent of marking you as having gone through a bankruptcy (which takes 7 years to clear).

Analysis Of Perfect Credit Score Math

Lets review the criterion for the ideal score. First, you must acquire as many lines of credit (unsecured debt) as you can. For each credit card, you must carry a monthly balance of 20% of your limit on each account and pay each bill on time. Assume a hypothetical case of a person who earns $100,000 and has credit cards with limits adding to $32,000.

Lets further assume this is a perfect score and you get an interest rate of prime + 8% which is essentially 8.5%. Carrying a 20% balance translates to always owing $6400. So the first cost of this status symbol is 8.5%*6400 = 544 per year in interest.

Assume you always use your card to pay for everything. Annual expenses add to $32,000 which means you agree to pay an additional $32,000 * 3% markup for everything you buy which translates to you agreeing to another $1000 per year in real cost so you can earn some lotto point called Rewards.

In this program you roll the dice to see how many cards you can maintain with this ratio. Banks know that if you are ever a day late in payments they can immediately raise your rates to 20 or 30% and hit you with a $35 fee. This is likely after you manage to outspend your income by using more that 20% of your limit due to some unexpected expense. Banks then win the Trifecta while you pay through the nose.

To counter the argument that the current prime rate does not allow banks to offer 5% savings accounts, thus far, even if you keep the perfect score, you are paying 1544 to the bank via transactions and the frozen balance of 20%. Any bank could easily afford to pay 5% interest on any amount up to 25,000 with profit left over.

As it stands, by simply paying off the 20% balance you pay yourself 8.5% interest on $6400. By paying in cash, you save $1544 per year in price adjustments and less costs.

The other upside of this method is when you pay with cash, you first save for purchases, then pay in full on the spot which also makes the funds immediately available to the merchants with whom you deal.

Real Savings Of Local Goods

When you walk into the big box grocery section, ever wonder how apples from Bolivia could ever really be on sale for 99 cents a pound in the city of Austin, Texas or Portland, Oregon. The honest is they are not that cheap and never really will be. Compared to a locally produced apple at $3 a pound the elements of cost are as follows:

1. Cost to procure from the grower that embodies their operational costs including labor and materials

2. Cost to transport to the market (in the case of a big box in US, this is sum of transport from farm to ship or plane+cost on plane or ship+transport via train and truck to big box warehouse+transport to big box)

3. Mark up by big box to sell it (to cover cost of store, employees and profit)

Careful examination of these elements reveals that we, the customers of the big box, look the other way on item 1, subsidize item 2 across the board by paying more somewhere else (sort of bait and switch) If actual accounting were in play, that apple from Bolivia would stack up as $9 a pound and we would sponsor the local grower.

Savings Based Economy

You have heard the mantra of save for the future. The health of our economy should be judged on the tradition measures of financial security not the securities and exchange commission view of which company is unlikely to fail NOW. To ensure growth, we need to demonstrate that we produce more than we consume. For a sustainable future we also need to see companies and individuals saving for the future and investing long term in capitol.

For individuals the single largest investment they make is in their home. Rewards should be granted for those who own a home and not simply refinance and drain any equity. Likewise, there should be more incentive for companies to hire local, build local and invest locally.

Using Cash Equates To Trust

The use of cash in transactions implies that immediate, full payment is being exchanged for a product. In this case there is no third party (credit card company or bank in the middle) The payment is immediately liquid with no need to wait on bank floats or transaction delays. There is no question that the payment has been made in full and is complete at bill of sale.

Contrast this with any form of credit or debit card transaction, the merchant must pay a fee for use of the service of processing cards. Admittedly with Square, less than before but still a fee. Then there is a percentage taken of the sale for each transaction. You may say, you pay "the same price" cash or credit but the reality is these fees result in higher prices for everyone so we all pay while only banks receive.

There is an expression of "hard earned cash". Whenever you offer someone cash in exchange for a product or service , you are using your capitol or liquid assets to pay. This constitutes immediate payment with no strings attached. Banks and credit card companies find this practice to be "unwise" and offer detriments to discourage such reckless behavior such as dings in your credit score due to behavior befitting money laundering.